THE POETRY OF ASTATINE

The Poetry of Astatine

Walter the Educator

Silent King Books

SILENT KING BOOKS

SKB

Copyright © 2024 by Walter the Educator

All rights reserved. No part of this book may be reproduced in any manner whatsoever without written permission except in the case of brief quotations embodied in critical articles and reviews.

First Printing, 2024

Disclaimer
This book is a literary work; poems are not about specific persons, locations, situations, and/or circumstances unless mentioned in a historical context. This book is for entertainment and informational purposes only. The author and publisher offer this information without warranties expressed or implied. No matter the grounds, neither the author nor the publisher will be accountable for any losses, injuries, or other damages caused by the reader's use of this book. The use of this book acknowledges an understanding and acceptance of this disclaimer.

"Earning a degree in chemistry changed my life!"
– Walter the Educator

dedicated to all the chemistry lovers, like myself, across the world

ASTATINE

Atoms dance their waltz,

ASTATINE

There lies a tale of Astatine's exalted salts.

ASTATINE

A fleeting phantom in the periodic scheme,

ASTATINE

Obscured in secrecy, a dreamer's dream.

ASTATINE

From halogen kin, it takes its stealthy stance,

ASTATINE

Rare and elusive, in atomic dance.

ASTATINE

A ghostly whisper in the cosmic choir,

ASTATINE

Astatine, the muse of scientific desire.

ASTATINE

With valence electrons in a fleeting array,

ASTATINE

It teases the senses in a cryptic display.

ASTATINE

An enigma cloaked in atomic attire,

ASTATINE

Astatine's allure, a scientist's fire.

ASTATINE

In hidden realms, where mysteries dwell,

ASTATINE

Astatine weaves its alchemical spell.

ASTATINE

In shadowy corners of the chemical stage,

ASTATINE

It dances with elements, an enigmatic sage.

ASTATINE

In isotopic guise, it morphs and transforms,

ASTATINE

Chasing its brethren in molecular swarms.

ASTATINE

A whispering phantom in the chemist's sight,

ASTATINE

Astatine beckons, in the depths of night.

ASTATINE

With radioactive glow and atomic charm,

ASTATINE

It captivates minds, casting shadows of alarm.

ASTATINE

A fleeting glimpse of the cosmic unknown,

ASTATINE

Astatine's secrets, to scientists are shown.

ASTATINE

In laboratories hushed, with beakers aglow,

ASTATINE

Astatine reveals its secrets, ever so slow.

ASTATINE

A tantalizing puzzle, a riddle untold,

ASTATINE

In its atomic dance, mysteries unfold.

ASTATINE

From Mendeleev's vision to the modern quest,

ASTATINE

Astatine persists, in its atomic unrest.

ASTATINE

A puzzle piece missing, yet to be found,

ASTATINE

In the alchemist's dream, it continues to astound.

ASTATINE

In chemical equations and theoretical lore,

ASTATINE

Astatine's presence is felt, forevermore.

ASTATINE

A shimmering mirage in the scientist's gaze,

ASTATINE

Astatine enchants, in its mysterious ways.

ASTATINE

ASTATINE

So let us raise a toast to this element rare,

ASTATINE

To Astatine's mysteries, beyond compare.

ASTATINE

In the tapestry of elements, it holds its place,

ASTATINE

Astatine's legacy, a cosmic embrace.

ASTATINE

ABOUT THE CREATOR

Walter the Educator is one of the pseudonyms for Walter Anderson. Formally educated in Chemistry, Business, and Education, he is an educator, an author, a diverse entrepreneur, and he is the son of a disabled war veteran. "Walter the Educator" shares his time between educating and creating. He holds interests and owns several creative projects that entertain, enlighten, enhance, and educate, hoping to inspire and motivate you.

Follow, find new works, and stay up to date
with Walter the Educator™
at WaltertheEducator.com

www.ingramcontent.com/pod-product-compliance
Lightning Source LLC
LaVergne TN
LVHW051921060526
838201LV00060B/4106